Making Treasures from Trash

Making Treasures from Trash

Marion Darman

Illustrations by
William Darman

South Brunswick and New York: A. S. Barnes and Co.
London: Thomas Yoseloff Ltd

© 1975 by A. S. Barnes and Co., Inc.

A. S. Barnes and Co., Inc.
Cranbury, New Jersey 08512

Thomas Yoseloff Ltd
108 New Bond Street
London W1Y OQX, England

Library of Congress Cataloging in Publication Data

Darman, Marion, 1937–
   Making treasures from trash.

   1. Handicraft.   2. Waste products.   I. Title.
TT157.D37   745.5   74–9280
ISBN 0-498-01609-9

Printed in the United States of America

To my "4 Js"

Contents

	Introduction	9
1	Bleach and Detergent Bottles	13
2	Play Clay	19
3	Cigar Box	24
4	Old Picture Frames	27
5	Stone Crafts	34
6	Old 45 RPM Records	38
7	Cardboard Tubes	40
8	Yarn Projects	43
9	Plastic Fruit and Vegetable Baskets	50
10	Papier-Maché	53
11	Plaster of Paris	58
12	Empty Cans	62
13	Milk Cartons	70
14	Pinecones	74
15	Old Greeting Cards	80
16	Egg Shell Crafts	84
17	Potpourri	89

Introduction

As you glance through the pages of this book, you are probably thinking, "Not another book on handicraft!" Yes, but this book has been written for all of you Arts and Crafts buffs, whether first-timers or lifers, who need new and refreshing ideas.

As a veteran of many P.T.A. fund-raising functions, and a two-year hitch as Den Mother for the Cub Scouts (with medals for bravery above and beyond the call of duty), I discovered countless ways to create attractive, useful and fun items from trash (that's right, trash).

No special skills or talents are required. You need not be a woodwork major or a graduate of an art-design institute. No blueprints have to be read and no special tools are required (it took me eleven years to recognize a Phillips screwdriver). All of these projects can be done by youngsters. There is, however, one requirement—you must become a collector. Nothing is ever discarded as trash until you have convinced yourself that it has no future use in any shape, size or form in any room of your home, or in your garage or backyard. You will also have to train your neighbors, friends and relatives not to put out their garbage for collection until you have personally checked for "valuables."

Whenever you leave the house it will become second nature to include a plastic or paper bag in your pocket, lest you discover a treasure of trash away from your neighborhood. Many a priceless bleach bottle has been left behind because my children forgot their "loot bags."

As the title implies, most of the items in this book are made entirely from throwaway items. Those that are made from purchased materials cost only pennies; all are priced under $1.00. Everything can be found at local merchants, so a 60-mile trip to an auto graveyard need not be made. No obsolete items used. I once anticipated making a chair out of old fashioned ice tongs—not only did I not find the ice-tongs, I couldn't even find the ice man!

There are no chapters entitled "Rainy Day Projects" because I have discovered that adults and children can be just as bored on sunny days. Nor are any holiday projects noted as such; who is to say that an Easter egg tree cannot be enjoyed on Labor Day?

The contents are grouped by main ingredients, so if you find you have accumulated 16 shoeboxes full of margarine container lids, you will be able to turn to the right page immediately, even while balancing the loaded shoeboxes.

Clear and simple instructions are given, along with photos and illustrations which are easy to follow.

Making Treasures from Trash

Bleach and Detergent Bottles

Any size, shape and color plastic bottle can be used in these projects so do not, I repeat, do not throw any out. If any of your friends or neighbors are seen discarding these prized objects, refuse to show them the treasures you have created. This will certainly change their minds about next week's trash collection.

Piggy Bank

Gallon or half-gallon bleach bottles make the best banks. These piggy banks make great gifts for children, especially when you have started their savings with a roll or two of pennies.

Materials needed:
  bleach bottle

14 MAKING TREASURES FROM TRASH

   felt scraps
· 4 small corks
   pipe cleaner
   white glue

Cut various size felt scraps into shapes of your choosing. Some ideas might include flowers, circles, diamond shapes, etc. Glue onto plastic bottle at random spots. Cover bottle cap with circle of felt cut the same size. Use magic marker to make small circles on felt for nostrils. With bottle handle on top, glue corks into place on bottom of bottle, spacing corks so they resemble feet. Cut two circles of felt for eyes and glue on sides of handle. Use magic marker for pupils. Poke small hole into bottom end of bottle (near top) and insert pipe cleaner. Curl end of pipe cleaner around finger to resemble tail. Cut coin slot into top of bank.

Curler Caddy

For all of us out there who are not blessed with naturally curly hair or the features to look stunning with straight hair, there is only one way to handle our handicap—curlers. A pretty hide-away can be made for storage of all those magical items we must use at shampoo time.

Materials needed:
  gallon bleach bottle
  piece of fabric 8" x 21"
  needle and thread
  shoe lace

Cut across bottle just below handle. Do not discard handle (see end of chapter for uses). Using ice pick or long nail, pierce holes $\frac{1}{2}$" from cut bottle edge. Space holes at even intervals of approximately 1". With right sides together, fold fabric so that short lengths meet. Seam short side (hand sew or machine stitch). Leave seam free 1" at one end. At this end, turn under raw edge $\frac{1}{2}$" and hem. Turn fabric right side out. Turn under remaining raw edge $\frac{1}{2}$" and fit fabric over bottle edge, having raw edges even with bottle edge. Thread needle with strong thread, knotting end well. Draw thread through fabric and hole until material is all firmly attached to bottle. Knot end of thread securely. Draw shoe lace through top hem to form drawstring closing.

Tissue Cover-up

A decorative cover-up for extra toilet tissue rolls. This is a good item for any fund-raising functions.

Materials needed:
  half-gallon bleach bottle
  plastic cafe curtain ring with eye
  plastic lace doily
  assorted "jewels" (pearls, sequins, etc.)
  white glue
  pipe cleaner or wire
  optional—model paint

Cut across bleach bottle $5\frac{1}{2}$" from bottom. Paint with model paint, if desired. Cut motifs from plastic lace doily and glue several into a

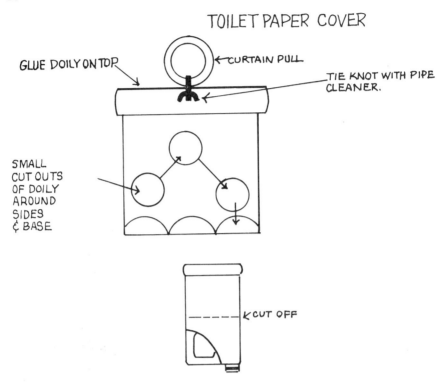

pattern at edge of inverted bottle (see illustration). Glue remaining doily cut-outs on sides and top of bottle, trimming with glued-on "jewels." Pierce hole in center top with ice pick or nail and insert eye of curtain ring. Secure eye inside with piece of pipe cleaner or wire. Place over tissue roll.

Left-over Bottle Pieces

Use left-over bleach bottle tops as funnels. Keep one or two in trunk of car for under-the-hood use (pouring motor oil, gas or water).

Salvaged pieces of plastic bottles may be cut into bubble-blowers for the children.

Mini-holder

Youngsters, especially, enjoy making these holders which can be used for countless items. One member of our family uses it to keep pencils in, while Mom uses her mini-holder for storing a scouring pad.

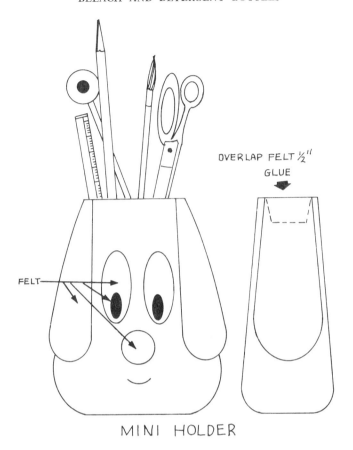

MINI HOLDER

Materials needed:
  dish detergent bottle
  felt scraps or wallcovering samples
  yarn ball (cut from ball fringe trim)
  white glue

Cut detergent bottle across approximately 4" or 5" from bottom. Cut 2 circles out of desired material and glue onto cut bottle for eyes. Glue yarn ball into position for nose. Cut "mouth" out of desired material and glue into position. Cut 2 ears as shown, fold down ½" and glue folded edge into inner edge of bottle.

Twine Holder

A useful "beehive" to store twine untangled and ready to use at a moment's notice.

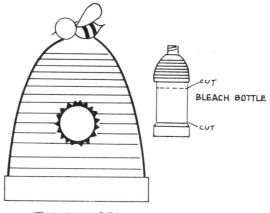

TWINE HOLDER

Materials needed:
quart bleach bottle
scrap of rickrack
yellow model paint
white glue
optional—tiny bees (available at florist)

Cut off tapered top of one-quart bleach bottle. Cut hole 1" in diameter in side, glue rickrack around hole. Cut bottom of bottle, leaving 1¾" sides. Paint both bottle pieces yellow. When dry, place twine into bottom half. Slip top portion over bottom and draw end of twine through hole. Glue bee onto bottle cap.

2
Play Clay

This project was initiated during chicken pox season, and having three children home at the same time in various stages of the pox, provided many hours of creativity.

PLAY CLAY BASIC RECIPE:

Mix in bowl until blended: 1 cup flour, ¼ cup salt and ¼ cup talcum powder. Gradually add ¼ to ½ cup water until mixture is soft, but not sticky. Knead mixture until completely smooth. If desired, food coloring may be added with the water.

Most pieces dry overnight, heavier ones take longer. When pieces are completely dry they may be painted with plastic paints, water colors or model paints. Pieces may be coated with clear shellac or clear nail polish.

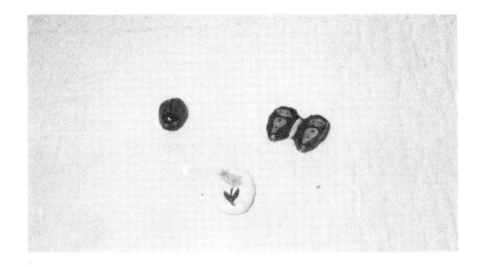

Play Clay Jewelry

Roll clay to ¼" thickness. Cut with cookie cutter or with small knife into desired shape. Re-roll to distort design or press in design with thumb. When making medallions or necklace, pierce small hole near top edge. When piece has dried, paint as desired. Suspend on a chain or length of narrow ribbon for necklace. For ring, press wire into piece before it has dried, add a bit of clay over the wire and smooth edges. For pin, glue pin clasp on back of dried piece. Glue "jewels" on painted pieces if desired. Use clear shellac or clear nail polish for protective coating.

Wall Decorations

Here's an opportunity to decorate the walls of a kitchen, den or child's room inexpensively. Enjoy a splash of color and ingenious design. For the kitchen, try a grouping of fruits and vegetables such as watermelon slice, orange and lemon slices, melon wedge, or try an onion, beets, carrots and squash (all easy patterns to cut). How about some gingerbread men? For a child's room try your hand at some animals (use a coloring book for ideas and patterns). Create an arrangement of alphabet blocks or use the letters of child's name. The ideas are endless. The signs of the Zodiac would be interesting for a den or play area. Try a large flower cut out for a mirror plaque.

Mirror Plaque

Materials needed:
  play clay
  paint
  small mirror
  white glue
  hair pin or short length of wire

Roll out clay to 1" thickness. Cut out flower shape approximately 6" in diameter. Press petals in with thumb. Cut out stem approximately

22 MAKING TREASURES FROM TRASH

1½" wide and 7" long. Cut 2 leaves each 2" by 4". Position stem, leaves and flower. Moisten small areas of clay where pieces are to be joined and press together. Position mirror into center of flower, making a depression. Remove mirror. Insert hanging hook made from bent hair pin into back of flower. When clay has completely dried, paint as desired. Glue mirror into depression. Use shellac or clear nail polish for protective coating.

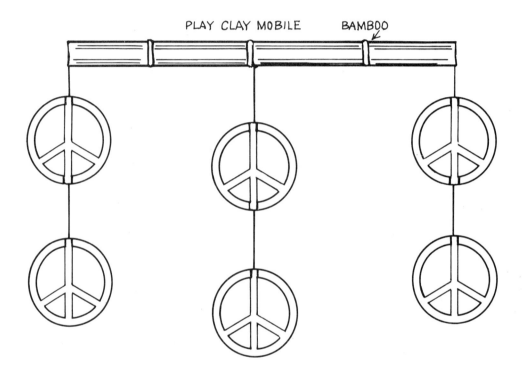

Play Clay Mobile

Materials needed:
  play clay
  cookie cutters
  paint

Roll out clay to ¼" thickness between pieces of waxed paper. Use cookie cutters to form shapes. Before clay dries, make small hole at top of form and 2 holes at bottom for hanging. Set aside to dry on wire racks (the kind used for cooling cakes). When dry, paint on both sides and insert thread through holes. Assemble mobile by joining forms with thread. Cut slits into ends of piece of bamboo pole and insert thread for hanging.

3
Cigar Boxes

Through the years, cigar boxes have been used as "catch-alls" by adults and children alike. They are compact, sturdy and come in various sizes and are usually kept in closets and drawers. However, after reading the following ideas, you will probably display them proudly (as well you should) for all to see and admire.

Jewelry Box

This finished product will appeal to men of all ages, so make several as gifts for all the males on your list. Ideal for the "man who has everything." I guarantee he won't have one of these!

Materials needed:
   cigar box
   paint
   clear varnish
   canceled stamps
   white glue
   felt scraps or wallcovering sample

Paint outside of cigar box desired color. When dry, glue on canceled stamps on all sides and top. Glue stamps in irregular pattern, overlapping stamps but leaving some color showing between stamps. After all stamps have been glued on, apply thin coat of varnish over entire box. Let dry thoroughly (overnight) and apply second coat of varnish. Allow to dry overnight again. Measure inside of box and cut felt or wallcovering to fit. Glue into place. Dividers may be made by cutting pieces of cardboard and covered with fabric to match inside of box. Glue into place at sides of box.

Memory Box

A sure hit with the Grandmas and Grandpas. The memory box

makes an ideal gift for Mother's Day or Father's Day, or make it an occasion by giving someone this gift.

Materials needed:
  cigar box
  woodstain paint
  white glue
  photo of your choice
  wallcovering sample (woodgrain)
  assorted trimmings

Paint outside of box with wood stain. When dry, line inside of box with woodgrain wallcovering, making sure any wrinkles are smoothed out before glue dries. Cut selected photo into oval or circular shape and glue onto center of box top. Outline photo edge with small piece of trimming such as rickrack, lace, ribbon, etc., glued into place. Include in memory box family photos, listing names, dates, occasions, etc.

These are but a few ideas of what can be done with cigar boxes. Almost anything can be used to cover and thereby disguise them. Use small buttons, shells, beads, even macaroni to create one of a kind items.

4
Old Picture Frames

During spring cleaning time, at garage sales and flea markets there is always an abundance of unusual picture frames, some with glass, mostly without. A variety of unique items can be made with these discards.

For those frames that still have glass intact, why not create a crushed glass mosaic or perhaps a gravel mosaic? For the crushed glass mosaic you will need various colored empty bottles and jars; beer bottles (brown) 7-Up bottles (green), cemetery vigil light (red), Noxzema jar (blue). Got the idea? For any possible color glass you feel like using, there is always someone who has an odd drinking glass (purple?) or cracked vase (opaque white) just waiting for you to find a use for.

The easiest and safest way I have found to break up these items is to place them, one at a time, between several thicknesses of newspaper, folding down and under all open ends. Use a firm surface such as a sidewalk to work on. With hammer, strike several blows until bottle is completely broken up. Carefully unwrap newspaper and without handling glass, pour fragments into suitable container (empty margarine cup). Use separate container for each color. Use tweezers to handle any crushed glass that has stuck to newspaper.

Crushed Glass Mosaic

Materials needed:
  picture frame with glass
  crushed glass
  aluminum foil
  household cement (clear)

Remove glass from frame and cover cardboard backing with foil. The foil side will show through glass, so glue edges so that no foil

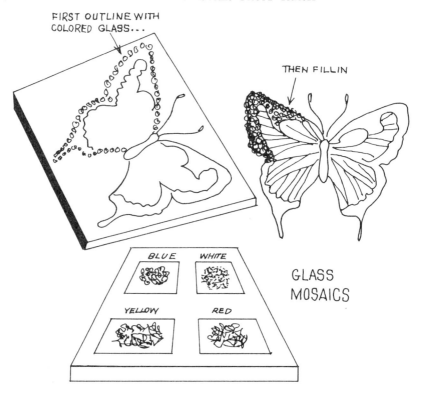

shows on hanging side of cardboard (neatness counts). If desired, crush foil slightly before gluing onto cardboard. Draw simple pattern onto foil and insert back into frame behind glass. Spread glue onto glass following outline drawn on foil. Work on small area at a time so that glue does not dry before glass is applied. Using tweezers, apply crushed glass onto glued areas until entire surface is covered. You may choose to cover entire glass surface for a stained effect or to leave some foil backing showing. Either method is very effective. When glue has completely dried, hang mosaic either singly or use in grouping.

Use multi-colored fish tank gravel in the same manner to make a gravel mosaic.

Dried Bean Art

For a nice touch in the kitchen, try **Dried Bean Art**.

Materials needed:
  picture frame with glass
  various types dried beans (lima, navy, pea, black eye, etc.)
  aluminum foil
  clear household cement
  clear lacquer
  optional—paint

Follow procedure for backing as given in crushed glass mosaic. On foil, draw desired picture (fruits, vegetables, cooking utensils, etc.) Apply cement to glass following drawn guide lines. Place beans into glued areas until pattern is completed. Leave natural or paint beans. Cover bean area with clear lacquer.

For all those frames that have no glass, here are several suggestions for getting them out of retirement. After trying these, I'm sure there will be many other ideas you will have for displaying these long forgotten frames.

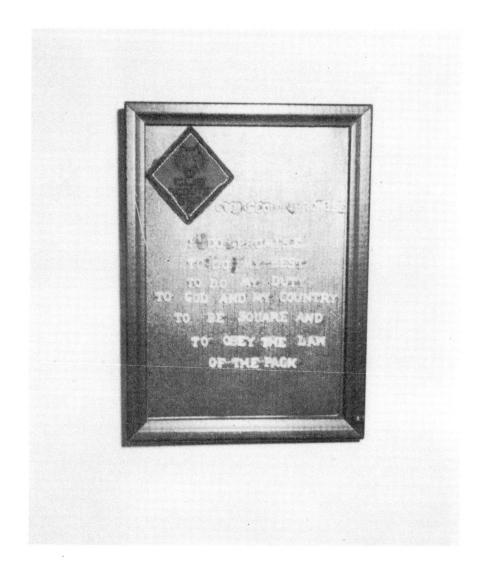

Motto Plaques

Children love doing these, and enjoy seeing their finished products displayed. This is a foolproof undertaking and the results are always rewarding.

Materials needed:
  picture frame—no glass
  alphabet macaroni
  white glue
  poster paints or water colors

Paint cardboard with bright color so macaroni will show up. When paint is dry, spread thin line of glue across writing line. Press "letters" into glue. Some ideas to follow: Child's Prayer ("Now I Lay Me Down To Sleep"), The Pledge of Allegiance, Boy Scout or Girl Scout pledge.

One of my youngsters made a motto plaque that read: "I made this plaque on Tuesday, November 11, 1972," then spelled out his name, address and age. It was the only one of its kind.

Calico Pictures

Materials needed:
  picture frame—no glass

fabric scraps
yarn scraps
white glue
page from coloring book

Choose a simple picture from child's coloring book. Use flower, animal, butterfly, etc. Glue page onto cardboard. Using different colored fabric scraps, cut out pattern pieces and glue onto picture. When picture is all fabric-covered, cut solid color fabric for background. Glue into place. Use yarn scraps to outline picture, gluing into place.

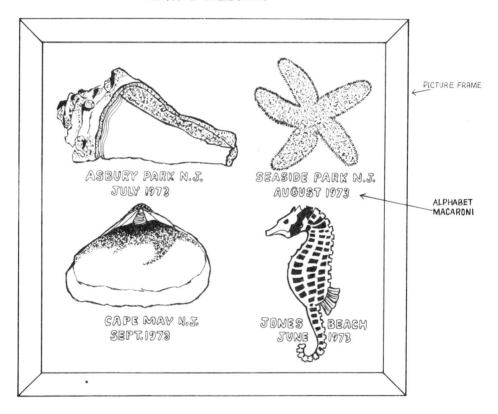

Treasure Collection

We have yet to return from a trip to the beach or park empty-handed. There are always so many priceless treasures that cannot be

Piggy Bank

Mirror Plaque

Party Favors

Stone Crafts—Rock Residents

left behind. These include tree bark samples, pebbles, shells and glass fragments worn smooth by the sea. To bring these valuables home and not be able to enjoy looking at them is unthinkable, so we have made treasure plaques.

Materials needed:
  picture frame—no glass
  aluminum foil or felt or paint
  white glue
  alphabet macaroni—optional

Cover cardboard with foil or felt, or paint desired color. When dry, glue on "treasures." Should memory fail as to when and where these treasures were discovered, alphabet macaroni can be used for identification. I even made one of these treasure collections using pennies I have found through the years. (I now boast a total of 8 pennies on my plaque).

5
Stone Crafts

These "rock residents" have been favorites wherever displayed. Your own imagination is your guideline in creating these lovable "citizens." My own collection includes a Girl Scout, Cub Scout, doctor, fireman, racing car driver, an elephant, teddy bear and even Humpty Dumpty. I have created "creatures" that have no known names or ancestry, but that have provided many a chuckle to anyone who views them.

Generally, the shape of the rock determines the ultimate character it will become. I have been known to spot an elephant body at 8 paces while strolling through the park. Parks provide a never-ending supply of every imaginable part of stone anatomy needed. Armed with empty shoe boxes, my family joins me on my rock hunt with each member assigned a different "body" item (Jeff, you look for heads—Jason, start collecting feet). We have drawn curious stares on more than one occasion from passersby who overhear one of us shouting our discovery of a panda body or a skunk tail.

These finds are collected and stored in boxes labeled "feet, heads, bodies," etc., until assembled and glued. I often wonder if the watermeter man ever notices them piled neatly on the basement floor, or does he normally run up the stairs three at a time?

Stone Crafts—Rock Residents

Materials needed:
  smooth rocks and stones of various shapes and sizes
  epoxy cement
  model paints

Arrange rocks to form desired "resident." Be sure base (feet) is level and that "head" sits evenly on body. Cement rocks into position and let dry approximately 8 hours. When epoxy has completely dried, paint rocks as desired. After paint has dried, draw features (faces,

STONE CRAFTS 35

arms and hands, etc.) with pencil and paint over using contrasting colors.

STONE CRAFT-FINE FEATHERED FRIENDS

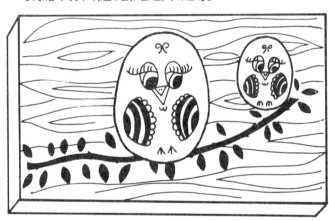

Fine "Feathered" Friends

These stone birds can be mounted on just about anything. Use wood scraps, cork squares, mirrors or old dinner plates. They can be hung on walls or stood on easel type stand.

Materials needed:
    smooth oval or round stone

model paints
household cement
small slender tree twig

Paint stone desired color on both sides. Allow to dry. Draw features on stone with pencil (follow illustration). Paint over pencil-drawn lines using various colors. When paint has completely dried, glue onto chosen backing. Glue twig into place so that bird is "perched" on it. Paint tiny leaves above and below twig. Try arranging 2 or 3 "birds" of different sizes on a branch.

6
Old 45 RPM Records

This project does not include any Enrico Caruso recordings, nor any collector's items (I am still holding onto them). After listening to three children's collections of assorted nursery rhymes over a period of 13 years, one is inclined to (1) buy large ear muffs, (2) feign deafness, (3) re-cycle the records. The solution was obvious since (1) people stared at me at the 4th of July cookout, (2) although the postman always rings twice, I missed receiving 2 packages, the Avon lady and the water-meter reader.

OLD 45 RPM RECORDS
Napkin Holder

Materials needed:
45 RPM record
paint
assorted trims

Preheat oven to 275 degrees. Place record directly on oven rack (one or two at a time) for 15 - 20 minutes. Using oven mitt, remove record from oven and quickly shape into holder by folding rounded edges in half with label portion in center on bottom, as illustrated. Work on flat surface so that base of holder (label portion) is level. Allow to cool before painting or decoration. For napkin holder, paint with model paints, spray enamel, etc. Glue any trims desired on sides of holder, or spread white glue on area and sprinkle with glitter.

This makes an interesting letter holder for Dad, too. After paint has dried, glue canceled stamps all over sides of holder. A real conversation piece!

Snack Plates

A clever way to serve snacks to teenagers after a football game or any sporting event, or at a slumber party (do they ever go to sleep?). First, try to use your teenager's most offensive (to only you, of course) rock records. Place in oven as for napkin holders. Remove from oven with oven mitt and quickly flute edges with fingers while molding record into bowl shape. Do not paint, simply line with paper doily before filling with snacks.

7
Cardboard Tubes

The cardboard tubes from rolls of paper towels, waxed paper, aluminum foil, etc., can be used for all sorts of year-round projects. They take their place proudly on the Christmas tree, at birthday parties, and at "Company" dinner parties.

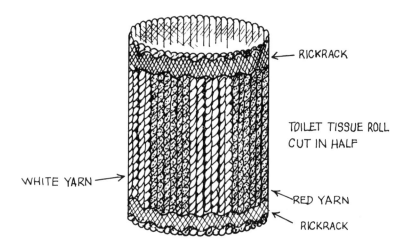

Drummer Boy Tree Ornaments

Materials needed:
  cardboard tubes
  yarn (red and white)
  gold rickrack
  white glue

Cut tube into 3" long sections. Wrap red and white yarn alternately through center and around tube, creating striped effect. Continue winding yarn until tube is completely covered. Secure yarn ends with glue on inside of tube. Glue gold rickrack over yarn in diagonal pattern as shown. Attach to tree with ornament hanger.

Party Favors

"Snappers" are a party favorite with children. I mention only children, as I am sure there are legions of mothers like me, who can recall the din caused by 20 snappers snapping simultaneously. The following "silent" snappers are just as pretty, contain the same goodies as the noisy ones and are a favorite of all my party-giving friends.

Materials needed:
  cardboard tubes
  crepe paper
  narrow curling ribbon
  assorted penny charms, balloons, gum balls, etc.

Cut cardboard tubes into 4" lengths. Cut crepe paper into pieces each 8" square. Wrap cut tube into square of crepe paper, securing one end with curling ribbon. Fill open end with goodies. Secure with curling ribbon. Attach small piece of paper with child's name, if desired.

Napkin Rings

Remember the good old days when every member of the family had his own napkin ring? Bring back fond memories, or start a new tradition with these up-to-date napkin rings.

Materials needed:
  cardboard tubes
  wallpaper samples
  small antiqued flowers (see potpourri)
  white glue
  small stapler

Measure and cut cardboard tube into 2" lengths. Cut wallpaper into 2½" by 7" strip. Glue over cardboard tube, bringing excess on edges down into underside of tube. Glue in place. Choose small antiqued flower and staple over seam of wallpaper.

8
Yarn Projects

If any readers do needlework of any kind, here is an opportunity to use all those left-over yarns. For those who do not knit, crochet or hook rugs, ask friends and relatives for their yarn scraps.

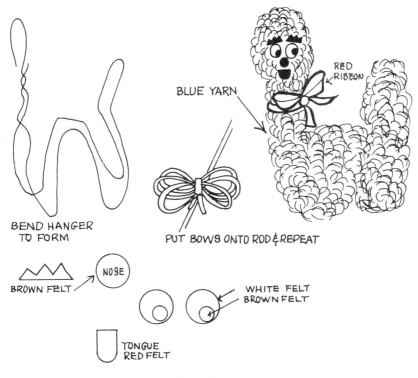

Yarn Poodle

A delightful addition to any child's room. Use true-to-life colors or match color scheme of room.

Materials needed:
  wire clothes hanger
  rug yarn (2 skeins)
  felt scraps
  white glue

44 MAKING TREASURES FROM TRASH

YARN PROJECTS 45

Bend wire hanger into basic poodle shape as illustrated, using rounded hanger end for head. Cut yarn into 3' lengths. Wind each length of yarn around three fingers, forming loops and tie in center. When all yarn has been wound and tied into loops, start attaching to wire frame at head portion. Continue tying loops onto wire frame until completely covered. From felt, cut 2 eyebrows, 2 eyes and a tongue, as illustrated. Glue felt pieces onto yarn loops forming head. Tie ribbon around neck of poodle.

Bottlecandle Holder

Materials needed:
  empty bottle
  assorted yarn
  scraps of trimmings
  white glue
  plastic margarine container lid

Starting at base of bottle, wrap yarn around bottle, gluing as you work for approximately 5". Wrap contrasting color yarn for next 3". Repeat with another color yarn until entire bottle is yarn-wrapped. Secure end of yarn with glue. Glue desired trim at color divisions and at bottle neck and base. Cut circle 3" in diameter out of plastic lid. Cut small circle in center of lid the same size as candle base. Insert candle into hole in lid and into top of bottle.

Yarn Duster

A great, lint-free duster—the brighter, the better. Almost makes dusting a pleasure.

Materials needed:
  wire clothes hanger
  Rug yarn (1 skein)

Bend hanger into duster shape, as illustrated. Wrap handle completely with yarn and knot end. Cut remaining yarn into 6" lengths.

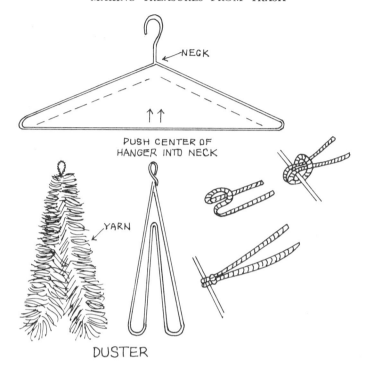

Fold each piece in half, drawing ends through loop onto hanger until entire wire is covered. These yarn dusters are great bazaar items. Happy housekeeping!

Yarn Ball Creatures

Great stocking stuffers or party favors. Make them by the boxful.

YARN PROJECTS 47

Materials needed:
 leftover yarn
 felt scraps
 lightweight cardboard
 white glue

Wrap yarn around fingers several times, forming loops. Tie in center. Cut through loops and fluff yarn into ball shape. Cut a piece of felt ½" by 1½". Cut a piece of felt into a ½" circle. Roll the longer piece and glue. Attach circle to end with glue. This is creature's nose, so glue to yarn at approximate center. Cut 2 small ovals of cardboard for eyes and with black crayon or magic marker, draw pupils. Glue to yarn above nose. Cut small oval of cardboard and use hole-punch or cut small hole in oval (palette). Dot with crayons to resemble paint colors. Glue to side of yarn ball. Cut oval of felt same size as palette and tiny strip of felt then glue strip to center of oval. This is creature's beret and is glued to top of yarn ball.

String Art

I have always used crochet yarn for this project, but several different weights of thread and yarn should produce interesting results.

Materials needed:
  piece of wood, any size and shape
  covering for wood piece (velvet, felt, aluminum foil)
  finishing nails or 3/4" brads
  crochet yarn

Cover entire piece of wood with one of the above choices. If using fabric, allow 1" more on all sides, bringing this allowance to back side of wood and gluing down. I have found that nailing brads at random points on front of covered board produces spectacular designs. If you would rather not be surprised, use a simple pattern such as a flower, leaf, etc., and nail brads according to pattern. If dark covering is used such as black velvet, light colored yarn shows up well. If using aluminum foil as covering, darker hued yarn gives best results. Nail brads into board leaving 1/4" exposed. Knot end of yarn on any brad in pattern and proceed to wind yarn from nail to nail, creating pattern as you wind. Wind yarn until a pattern is visible and repeat winding procedure until desired depth is reached (generally 10 to 12 windings). Knot end of yarn at last nail wound, cutting excess close to knot. Hang on wall and stand back to admire your artistic creation.

Crochet Yarn Eggs

This idea was originally thought of for Easter time, but has become a popular year-round activity. To suit your whim, you need only to decide what shape balloon to use.

Materials needed:
  crochet yarn
  liquid starch
  balloons

Blow up balloon to desired size and shape. Knot end of balloon. Pour starch from bottle into a bowl large enough to hold ball of crochet yarn. Submerge entire ball into starch. With wet end of yarn, start winding around balloon in all-over pattern until balloon is barely visible. The yarn should resemble several overlapping spider webs. Cut yarn, tucking end under wound yarn. Hang balloon up to dry by neck (use clothesline with paper under it to catch drips). Dry overnight. Remove from line and puncture balloon, drawing it out through yarn opening. You may choose to leave "eggs" as is, or cut small oval into side of egg and fill with Easter basket "grass" and small fuzzy chicks. Several lace eggs can be put into a basket for an unusual table centerpiece, or attach thread to top of eggs and hang mobile fashion in a window. Make round ones for the Christmas tree, too. The uses for these lace creations are unlimited.

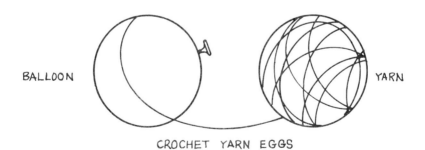

CROCHET YARN EGGS

9

Plastic Fruit and Vegetable Baskets

The plastic baskets that strawberries come packed in are ideal "whatnot" containers. Here are but a few ways of dressing up these throwaways.

Ribbon Holder

Materials needed:
 plastic fruit basket
 narrow ribbon

cardboard
white glue

Weave ribbon through basket openings and tie ends in small bow. Continue weaving one row at a time, tying bow at different place each time. Cut a square of cardboard to fit top of basket and glue strips of ribbon over cardboard. Glue a small bow onto center of top. A charming way to store hair ribbons and bows for the girls around the house.

Posie Basket

Materials needed:
 plastic basket
 wallcovering samples
 pipecleaners
 white glue

Follow directions given for making Gingham Garden Flowers, using wallcovering in place of fabric. Cut narrow strips of wallcovering and weave through holes of basket, gluing ends together. Use clay to secure posies in basket.

Button Basket

Materials needed:
 plastic basket
 wallcovering sample
 white glue
 assorted buttons
 cardboard
 empty thread spool

Cut wallcovering to fit inside of basket and glue into place. Glue assorted buttons on outside of basket on all sides. Cut square of cardboard to fit top of basket and cover with wallcovering. Glue spool into center of top for easy removal.

10

Papier-Maché

This magical stuff has always been one of my favorites. Imagination is the key ingredient. Papier-maché transforms ordinarily discarded items into exotic conversation pieces and attractive, useful articles. Making papier-maché is very simple, but messy. The small fry always like joining the activities because they can get messy without mommy's disapproval (she looks just as messy).

Cover your entire work area and immediate vicinity with lots of newspaper. Remove any jewelry from hands as the mixture is difficult to remove when dry.

Papier-Maché

Materials needed:
  newspaper
  flour
  water

Use a medium-sized plastic bucket or pail. Fill $\frac{1}{3}$-full with water. Gradually add flour a handful at a time, mixing with free hand to eliminate lumps, until mixture is consistency of thick soup. Tear (do not cut) newspaper into strips approximately 2" wide and 5" long. Work with 3 or 4 newspaper pages at a time so strips will be fairly even. That's all there is to it. Now let's create.

Apple Pencil Holder

A delightful way to keep pencils handy.

Materials needed:
  paste
  newspaper strips

54 MAKING TREASURES FROM TRASH

plastic apple
enamel paint

Make 3 or 4 holes in top of plastic apple. Use scissors to shape round holes just large enough to insert pencils. Tear newspaper strips into small squares then draw through paste and apply to plastic apple, overlapping pieces. Cover entire surface then repeat procedure until 4 layers are applied. Carefully insert pencils through wet papier-maché into holes cut into apple. Withdraw pencils (any ragged edges may be trimmed when dry). When papier-maché is completely dry, paint with red enamel. Paint 2 green leaves on top of apple.

Ornamental Wastebasket

Materials needed:
  paste
  newspaper strips
  twine or similar string
  cardboard box (any size desired) for wastebasket

enamel paint
optional—varnish

Draw newspaper strips through paste one at a time and wrap around box until completely covered, smoothing strips as you apply them. Inside of box may be covered if desired. Continue applying strips until 3 or 4 layers completely cover the box. Now, use your imagination and either (1) make irregular lumpy shapes with several pasty strips, or (2) draw twine through paste to apply to covered box. A combination of lumps and twine is interesting. When papier-maché has dried completely (allow 1 day), paint with spray enamel. Apply 2 coats of paint for full coverage, permitting first coat to dry before applying second coat. Varnish, if desired.

Party Pinata

Here's another delightful idea. Any party will be more exciting with a pinata. The children love it, and all too often, the adults have a crack at it too.

Materials needed:
 paste
 newspaper strips
 large balloon
 paint
 assorted favors

The shape of the balloon when blown up will roughly give the idea for the pinata (oval balloon suggests a football or a fish, round balloon could be used for a baseball or basketball, a smiling face, sun, etc.). Draw strips through paste and cover balloon entirely, applying 3 or 4 layers. Smooth strips as you work to avoid any wrinkles or lumps. Hang on clothesline by neck of balloon to dry approximately 24 hours. When completely dry, burst balloon with pin and withdraw. Carefully cut small semicircle into area where balloon was withdrawn. This opening is used to fill the pinata with assorted party favors. Tape opening shut. Paint with appropriate color (poster paints are ideal and quick drying). Use brown for football or basketball. Paint details such as lacing with contrasting color paint. Make 2 small holes on top of finished pinata and draw length of twine through for hanging on ceiling. When party time comes, blindfold one at a time, hand them a yardstick with which to break the pinata. Guests scramble for favors when pinata is broken.

PARTY PINATA

PAPIER-MACHÉ

Doorstop

Materials needed:
  brick, any size
  paste
  newspaper strips
  twine or yarn
  paint

Draw strips through paste one at a time and wrap around brick. Cover entire brick using 3 or 4 layers of strips. Draw twine through paste and swirl onto covered brick in any way you wish. Set brick aside to dry. Rotate brick periodically during drying time so every surface is exposed to air for quicker drying. When thoroughly dry, paint as desired.

This doorstop is an attractive way to prevent paper plates, napkins, etc. from blowing away at cook-outs.

11
Plaster of Paris

Another amazing material that has countless possibilities. Following are only a few of the many things that can be done with plaster of paris. As you read through these ideas, you will undoubtedly think of dozens more. A ten-pound bag of plaster costs under a dollar and can be used for several items.

Wall Plaques

Materials needed:
  plaster of paris
  water
  margarine container lid
  hairpin or similar type wire
  fishtank gravel
  white glue

Mix plaster with water in plastic bucket or bowl. Use 2 parts plaster to 1 part water, stirring constantly until any lumps are dissolved. Plaster hardens quickly (5 to 7 minutes), so work quickly. Pour mixture into lightly greased container lid, making sure mixture is even with lid edge. Bend hairpin or wire for hanging hook (see illustration) and insert into wet plaster near edge. Set aside to dry completely. This will take approximately one hour. When plaster has hardened, turn upside down and gently ease off plastic lid. On front side of hardened mold, draw a design with pencil. Spread thin coat of white glue in pencil outline and sprinkle with colored fishtank gravel. When glue has dried, excess gravel can be removed by holding plaque sideways.

An interesting arrangement can be made by using each of the card suits as center patterns.

Recipe Holder

A useful and unique addition to your kitchen is the following recipe holder. A good grab bag gift.

Materials needed:
  plaster of paris
  water
  2 ounce size plastic medicine cup
  plastic clothespin, clip type
  small plastic flower
  green poster paint—optional

Mix 2 parts plaster with 1 part water in small bowl, stirring to dissolve any lumps. Quickly pour mixture into medicine cup ¾ full. Before plaster sets, insert stem of small plastic flower into center of plaster. Insert one long end of clothespin into plaster at edge of cup, having other long end of pin on outside of cup, as illustrated. Set "flower pot" aside to dry. When plaster has hardened, paint top green, if desired. This makes a novel hostess gift when accompanied with a favorite recipe along with the finished product.

60 MAKING TREASURES FROM TRASH

Memo Holder

A memo holder for any golfer can be made in a similar way.

Materials needed:
  plaster of paris
  water
  2-ounce plastic medicine cup
  plastic clothespin, clip type
  plastic golf tee
  plastic toy golf ball

Follow directions for mixing and pouring plaster for recipe holder. Insert tee into center of plaster. Insert one long end of clothespin into plaster at edge of cup, having other long end of pin on outside of cup. When plaster has hardened, glue golf ball onto tee. Paint plaster surface green, if desired.

Lucky Paperweight

No game of chance involved here. These "dice" make unique decorations and can be used as paperweights, pencil holders or candle holders.

Materials needed:
  plaster of paris
  water
  pint-size milk carton
  black model paint

Cut top portion off a pint-size carton so remaining part forms a perfect square. Mix 2 parts plaster with 1 part water in plastic bowl, stirring until mixture is smooth. Pour mixture into carton, level with top edge. Use flat edge knife to smooth top surface. Set aside to harden. Carefully tear away carton from hardened mixture. Paint dots on all sides of square to resemble dice.

To make pencil holder, pour mixture into carton and level with knife. Before mixture hardens, lightly grease pencils (use cooking oil, shortening, etc.) and insert into top surface. Remove pencils when mixture has hardened and paint black dots on all sides as on dice.

To make candle holder, pour mixture into carton, level top surface with knife and insert candle. Remove candle from set mixture and paint dots on all sides as on dice.

12
Empty Cans

Ecology gets a big boost when empty cans are recycled into pretty and imaginative items for personal use. A few examples of what can be done with throwaway items are given. My favorite material for covering cans is wallcovering, either vinyl or paper. No, it is not necessary to redecorate your bedroom in order to have the wallcovering needed for these projects. Any wallcovering dealer will gladly give you sample books of discontinued patterns. These books contain hundreds of wallcovering swatches, many of which are color-coordinated. An attractive desk set can be made quickly and easily by using ordinary empty cans. Match the decor of the room or choose something wild and offbeat.

Desk Set

Materials needed:
  frozen juice can
  shortening can
  cardboard, cut from carton
  wallcovering samples
  white glue
  trimming scraps
  optional—small memo pad

To cover cans, cut wallcovering to measurement of can height and can diameter. Add 1″ to diameter measurement for overlap. Apply thin coat of white glue to back of wallcovering. Cover can with wallcovering, smoothing any wrinkles as you work. Overlap edge. If desired, trimming may be glued at top or bottom of covered can. Cut cardboard to exact measurement of wallcovering sample being used. Apply thin coat of glue to cardboard and smooth wallcovering over entire area. Glue trim on all edges. A small memo pad may be covered in same manner to complete desk set.

Crayon Box

If there are any youngsters in your home who belong to the Coloring Book Brigade, you know how long the original crayon box stays intact. To make an indestructable, almost impossible-to-lose crayon container, all you need is an empty coffee can or similar type can with plastic lid.

Materials needed:
  coffee can with plastic lid
  wallcovering sample
  white glue
  optional—magic marker

Cut wallcovering to measurement of can, adding an inch to length for overlap. Apply thin coat of glue to back of wallcovering and cover can, overlapping edge. If desired, use magic marker to write child's name on lid, or draw a crayon on lid for easy identification of contents.

Similar cans can be used for storage of small items such as puzzle pieces, marbles, hair barrettes, etc. Identify contents with drawings of items on lid.

Kitchen Cannister Set

An inexpensive (almost free) cannister set can be made from various sized cans. In addition to the usual cannisters for flour, sugar, tea and coffee, why not make matching ones for instant coffee, cocoa, crackers, etc? Use cans with plastic lids such as coffee cans, shortening cans, nut cans. These items come in assorted sizes, making it possible to have a graduated cannister set. Follow directions for making crayon container, using wallcovering in a kitchen print.

Egg Shell Flowers

Blooming Button Pot

Gingham Garden Flowers

EMPTY CANS 65

Clutter Cannisters

So-called because that's what I had all over the house before I organized. One never knew where to find a button, safety pin or thumb tack, when needed. Sound familiar?

Materials needed:
 cans with plastic lids
 wallcovering samples
 white glue

Use bright, cheery patterns of wallcovering, cut to measurement of can plus an inch for diameter overlap. Apply thin coat of glue to back of wallcovering and cover can. Glue sample of cannister contents to lid top, i.e., button from button box, safety pin from pin can, etc. Use woodgrain pattern wallcovering for cannisters for the man of the house. Glue samples of contents on lid, using nails, screws, washers, etc.

You will be amazed at how organized the family will become. No longer will you have to interrupt your pedicure (mid-toe) to find an orange button for Suzy or a $3/8''$ washer for Junior. Now, if you can only remember where you put all those clutter cannisters . . .

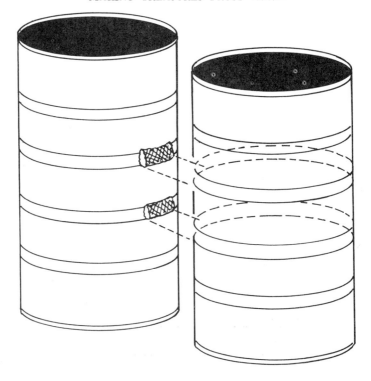

Wine Rack

Materials needed:
  46-oz. size juice cans
  rubber cement
  high gloss enamel paint
  rope, clothes-line type

Remove both ends of juice cans with can opener. Peel off labels, then wash and dry cans. Paint inside and outside of cans. Allow to dry. Cut rope into small pieces and glue into grooves on sides of can (see illustrations). Join 4 cans side by side for base, then join 3 or 4 layers of cans on top of base. Glue rope to can openings on front of wine rack. Can also be used as a magazine rack.

Card Holder

Materials needed:
  one 46-oz. juice can

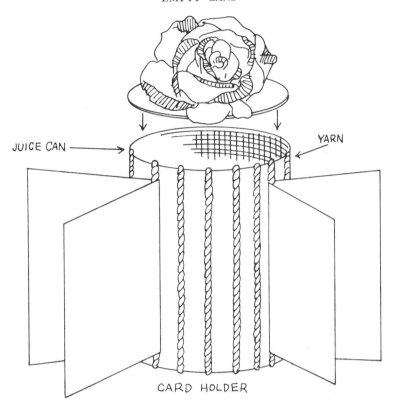

aluminum foil
rubber cement
red yarn
artificial holly

Remove both ends of juice can with can opener. Peel off label, then wash and dry can and one lid. Apply cement to outside of can and lid and cover completely with foil. Trim away excess. Tape one end of yarn inside can, then wrap yarn up and down on inside and outside, leaving approximately ½" spacing between strands. Tape loose end of yarn inside can. Adjust yarn so it is evenly spaced. Glue artificial holly on covered lid and push lid slightly into one end of can. Loop cards through yarn.

Tom-Tom

Let your favorite "warriors" try this just-for-fun tom-tom.

Materials needed:
  coffee can with plastic lid
  paint
  yarn scrap
  small wooden beads, macaroni, feathers, etc.
  white glue

Paint outside of coffee can and allow to dry. Apply thin strip of white glue to can diagonally from top to bottom then press yarn over glue. Slip yarn through small wooden bead or macaroni before gluing to can. Glue clusters of small feathers or paint Indian symbols on can. Use hand on plastic lid to send messages or make a tom-tom beater. To make beater, insert pencil into hole of empty wooden spool and paint spool to match can.

Cigarette Box

Materials needed:
  small vegetable or tomato sauce can

white glue
empty cigarette packs
clear varnish

Remove label, then wash and dry can. Cut front of cigarette packages into squares and glue around can, overlapping edges in an irregular pattern until can is entirely covered. Apply one or two coats of clear varnish to seal.

13

Milk Cartons

While the sanitation engineers are still very puzzled at the lack of garbage on your block, here are some ideas for using milk cartons. I'll bet you thought they were only good for disposing of leaky garbage!

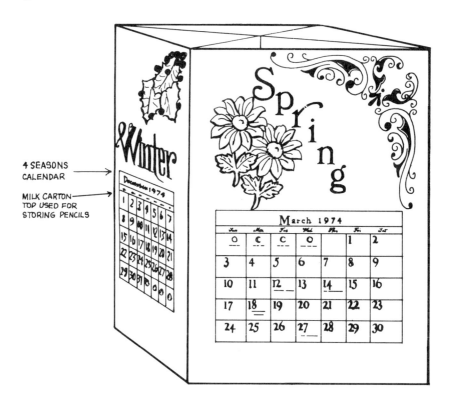

4-Season Calendar

Materials needed:
  milk carton

wallcovering sample or construction paper
white glue
small calendars (banks give them away)

Cut carton across fold where pouring spout is located. Glue wallcovering or construction paper to cover entire carton. Divide calendar into 4 seasons and glue on 4 sides of carton. Over appropriate season either write season (Spring, Winter, Summer, Fall) or draw picture of season's flower (use rose for Summer, tulip for Spring, holly for Winter, mums for Fall). Open end of carton may be used as base of calendar or turned upright for use as pencil holder.

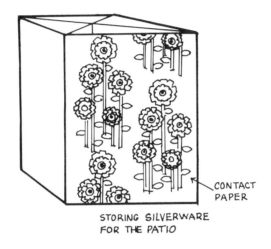

STORING SILVERWARE FOR THE PATIO

Plastic Cutlery Container

Remember the juggling act you performed the last time the crowd was over for a cookout? What with balancing paper plates, cups, napkins and cutlery, an octopus was to be envied. An attractive cutlery container can be made easily and provides an idea for various other containers.

Materials needed:
  empty milk cartons (2)
  wallcovering samples
  white glue

Cut down top of milk carton to fold where spout is located. Cut another carton into 2 sections to fit into first carton. These sections

72   MAKING TREASURES FROM TRASH

form dividers for cutlery. Cover entire carton with wallcovering. Divider sections may also be covered. Now it will be possible to carry knives, forks, spoons and serving utensils in one handy container, leaving other hand free for plates, cups, etc.

A popular pastime these days is candle-making. An inexpensive way to create pretty and unusual candles is to use discarded milk cartons in various sizes. One of my favorite methods is for making lacy candles.

Lacy Candles

Materials needed:
  milk carton
  paraffin wax (sold in supermarket)
  wax crayons (those which have been broken and discarded)
  piece of twine (for wick)
  crushed ice

Cut carton down from top to approximately 6" in height. Place paraffin (1 lb.) into empty coffee can. Place coffee can into saucepan containing water. Heat on stove, using small flame until wax has completely melted in coffee can. Add crayon from which paper has been removed to melted wax. Interesting effects can be achieved by melting several different colored crayons at once. When crayon has completely melted, remove saucepan from heat, using pot holder to handle coffee can. Place crushed ice into milk carton 3/4 full. Over ice, slowly pour melted wax. While still liquid, place length of twine into center, making sure twine goes almost to bottom of carton. Set aside to harden. May be placed in refrigerator to speed hardening. When wax is completely hardened, slowly tear carton away from wax. The smaller the milk carton, the less paraffin needed. Use approximately 1 lb. for a half-gallon carton. Several different sized candles make an interesting table centerpiece.

14

Pinecones

Every project I have ever run across using pinecones has been a Christmas season undertaking. I personally like the looks of pinecones all year round, and enjoy seeing them displayed in various ways.

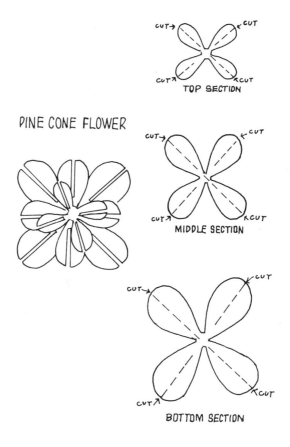

Pinecone Flowers

Flowers made from pinecones have a beautifully woodsy look.

PINECONES 75

They can be used for all season arrangements, left in their natural state or spray-painted.

Materials needed:
  pinecones
  fine wire
  optional—spray paint

Cut pinecone into 3 or 4 sections, using serrated-edge knife. Use top sections for flower buds. Using small scissors, cut edges of cones to resemble petals. Make one or two cuts per edge. Paint, if desired. Hook wire and insert between petals. Twist ends together. Arrange in small baskets or in a large container for table centerpiece.

Pinecone Owl

Materials needed:
  one whole pinecone
  2 pinecone sections

76 MAKING TREASURES FROM TRASH

   household cement
   2 marbles
   small tree twig

Using serrated-edged knife, cut 2 sections from pinecone. They should both be the same size. Glue marble into the center of each section. With flat, broad end of pinecone at top, glue "eyes" onto side of pinecone. Apply glue to small twig and press into place between petals at tapered edge of pinecone.

Pinecone Doll

Materials needed:
  pinecone
  acorn
  dried flowers
  pipe cleaner
  fabric scrap
  burlap scrap

PINECONES 77

Cut burlap into a 5" circle. Cut one slit into circle to center. Cut out small hole in center. Insert tapered end of pinecone into slit so that small bottom of cone is below center cut-out. Overlap edges of burlap and glue in place (burlap should not lay flat). Cut small square of fabric approximately 3" and glue top edge to burlap under pinecone petals. Cut small triangle of fabric and glue to acorn to resemble kerchief. Glue acorn to flat end of pinecone. Insert pipecleaner between petals for arms. Bend pipecleaner ends into small loops for hands. Place several tiny dried flowers into one hand.

Pinecone Centerpiece

Materials needed:
  3 large pinecones
  dried snowflowers (available at florist, variety store)
  green felt
  white glue

Using scissors, cut snowflower stems off at base of flower head. Using small dot of glue, attach one snowflower to each "petal" of pinecone. Repeat until all three pinecones have a snowflower on each petal (see illustration). Using a leaf pattern, cut 3 leaves from green felt. Join leaves together at widest point of each with small amount of glue. Glue one pinecone to each leaf at wide edge.

PINECONE BIRD FEEDER
PEANUT BUTTER AND BIRD SEED

Pinecone Birdfeeder

A good project for the smallest members of any group.

Materials needed:
  pinecones
  peanut butter
  wild birdseed
  thread for hanging

Using a popsicle stick or similar size utensil, spread even coat of peanut butter between "petals" of pinecone. Dip pinecone into wild birdseed until peanut butter is covered. Slip thread through pinecone "petals" at flat edge and attach to tree.

Pinecone Turkeys

These turkeys make unusual place settings for Thanksgiving get-

PINECONES 79

togethers. Grouped in the center of a table, they make a unique conversation piece for any Fall season gathering.

Materials needed:
  pinecones
  felt scraps
  pipe cleaners
  white glue

Cut pipe cleaner in half and bend to resemble turkey feet (see illustration). Glue long end of pipe cleaner into spaces between petals on pinecone near broad, flat end. Cut felt scraps into tail feathers, as shown. Glue into place between petals at tapered end of pinecone. Graduate width of felt feathers so that smallest one is at very end. Cut 2 scraps of felt into shape shown. Insert a short length of pipe cleaner between the 2 pieces, leaving approximately 1" of pipe-cleaner showing below felt. Glue both pieces of felt and pipe cleaner together. Insert exposed end of pipe cleaner into petals of pinecone at flat end. Glue into position. Eyes can be made with magic marker or by gluing small sequins on each side of "head."

15

Old Greeting Cards

No doubt you have now discovered many items that were previously considered trash. All those greeting cards you meant to throw away, but never did, can now be put to good use.

Photo Greeting Card

Materials needed:
  greeting card
  photo
  white glue

Fold construction paper in half across width. Choose greeting card with simple picture. Carefully cut around picture desired and glue to top portion of construction paper, as illustrated. Cut 4 small slits in top half of construction paper, next to greeting card cut-out. Insert photo. Open construction paper card and write personal greetings. Makes a nice remembrance for any occasion. I have used these cards as thank-you notes for children's parties, using photos of guests and animal cut-outs from old birthday cards.

Party "Loot" Bags

Materials needed:
  small lunch bags
  children's greeting cards
  white glue

Cut out pictures from greeting cards and glue onto bags. Write party guest name on bag and fill with "take-home goodies."

GREETING CARD PUZZLE

Greeting Card Puzzles

Children as well as adults enjoy these puzzles, simply made from old greeting cards. Cut out several pictures from greeting cards in various puzzle piece shapes. Place in envelope, using 2 or 3 different pictures. This makes a good pastime for children confined to bed. For a unique birthday greeting, cut up a new card (include the greeting portion as well as picture) and place in envelope for mailing.

Easy Decoupage

Materials needed:
  greeting cards
  white glue
  wood pieces
  sandpaper
  clear shellac

Use sandpaper to smooth top surface and all edges of wood. Cut out picture from greeting card and center on wood. Glue into place. When glue has dried, cover entire area with shellac. When dry, repeat shellac process until 5 coats of shellac are applied, allowing each application to dry thoroughly between coats. If desired, wood may be stained before greeting card picture is glued in place.

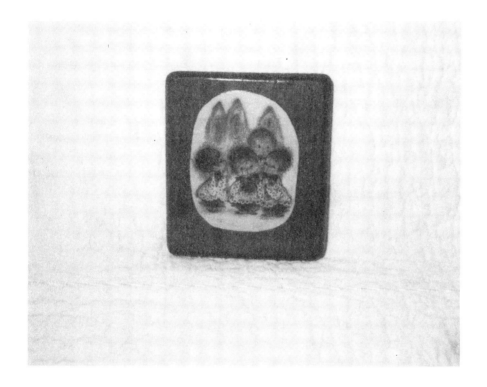

16
Egg Shell Crafts

The ideal way to accumulate enough egg shells for these items is to bake all the cakes for a P.T.A. cake sale (you'll be the hit of the school), or invite the neighborhood for Sunday brunch at which you serve omelets (you'll be the hit of the block). You can, of course, save all the egg shells you use daily until you have accumulated enough for these projects.

EGG SHELL ART — CRUSHED EGG SHELLS

Egg Shell Art

Materials needed:
  clean, crushed egg shells
  paint, poster or water color
  white glue
  art canvas (bought in variety store)

Either draw or trace simple subject (flowers, bowl of fruit, etc.) on canvas. Decide which areas are to be egg-shelled. If using flowers as subjects, paint stems and leaves and use egg shells for flowers. When painted areas have dried, apply glue to canvas where flowers are to be. Cover glue with crushed egg shells. If layered look is desired, allow first layer of glued shells to dry and repeat procedure until desired effect is obtained. When glued shells have completely dried, paint "flowers" desired color.

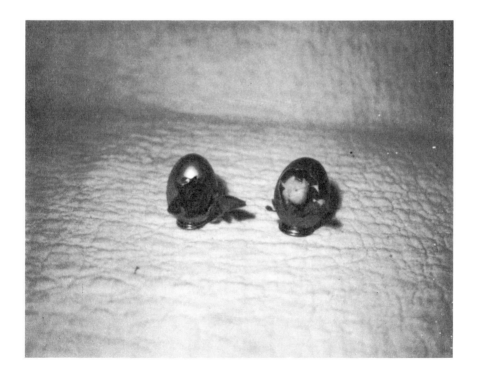

Golden Egg

A delicate addition to a knick-knack shelf, this golden egg makes a lovely gift at Easter time.

Materials needed:
  whole egg shell
  gold model paint
  gold café curtain ring
  small artificial flowers
  white glue

Before using egg for cooking, carefully poke hole into side (use straight pin). Slowly work pin around hole area, chipping small fragments out of shell. Finished hole should be only large enough to put one finger through. Carefully remove inner membrane from egg and wash inside of shell thoroughly. When dry, paint inside of egg and outer shell with gold model paint. After paint has dried, glue café curtain ring to base of egg. This enables egg to stand on shelf. Insert tiny flowers into small amount of modeling clay, and glue into place at base of inside shell. Glitter may be sprinkled on clay to which thin coat of glue has been applied.

EASTER EGG TREE

Egg Tree

Our family egg tree started out as an Easter table centerpiece and was so pretty to look at, it remained a centerpiece until Halloween.

Materials needed:
  whole egg shells (hollow)
  egg dyes or food coloring
  small plastic flowers
  assorted colored glitter
  white glue
  small tree branch with leaves removed

Before using eggs for cooking, pierce small hole with straight pin in both ends of egg. Blow egg out into bowl. Rinse egg shells thoroughly. To color hollow egg shells, place into dye solution and hold submerged with light pressure until desired shade is obtained. When eggs are dry, apply small amount of glue where glitter is desired, and sprinkle with glitter. Shake excess glitter off shells when glue has dried. Form "tree" by placing branch into shallow container. Use modeling clay to secure base of branch. Glue plastic flowers onto branch at random points. Glue short length of thread onto one end of egg (where hole was made). Cover with glitter. Hang eggs onto tree with thread. Whether you have a green thumb or not, your egg tree will bloom indefinitely and will be the envy of all gardeners.

Egg Shell Flowers

Materials needed:
  half egg shells
  pipe cleaners
  colored tissue paper
  white glue
  egg dye or food coloring

Color egg shell halves by submerging into egg dye. Dry thoroughly and pierce small hole into end of shell. Wrap pipe cleaner around a pencil to form a spiral at one end. Insert straight end of pipe cleaner in hole and carefully pull through until only spiral is within egg shell. Cut colored tissue paper into small squares and glue one edge into shell around spiral. Stick pipe cleaner stems into small amount of clay.

An ideal "vase" for these flowers is a spray-can cap. If decorated vase is desired, crush colored egg shells and glue around sides of cap.

EGG SHELL FLOWERS

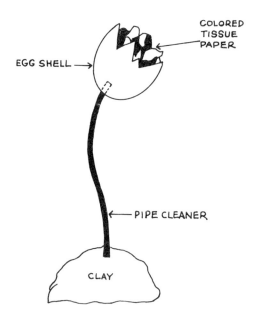

17
Potpourri

CUP AND SAUCER PLANTER
SHOE BUFFER
ANTIQUED FLOWERS
BLOOMING BUTTON POT
SPRAY-CAN COVERS
BAMBOO WALL HANGINGS
STAMP BOXES
DRAPERY RING WALL ACCENT
FAMILY TREE
CLOTHESPIN TRIVET
SPICE SACHET BALL
MINI-VASES
KEY JEWELRY
GINGHAM GARDEN FLOWERS
DECORATED SOAP
CORK TRIVET
FAMILY NAME PLATES
SPOOL HEADS

Potpourri

These items have been consolidated into one group because I have given only single examples of their uses. My reason for doing so is that after reading the previous chapters, many of you will undoubtedly start compiling your own books on other ways of transforming trash into treasures. Start sorting through all those accumulated, seemingly useless items in your closets, attics or cellars and enjoy using them to create the objects described in this section.

TURN SAUCER UPSIDE DOWN
GLUE CUP ONTO SAUCER
GLUE FRINGE & FELT FOR EYES,
SEQUINS FOR MIDDLE OF EYES,
& FELT FOR MOUTH, & WOOL
FOR HAIR.

Cup and Saucer Planter

Materials needed:
  cup and saucer
  rickrack
  2 small buttons or sequins
  yarn scraps
  white glue
  felt scraps

Invert saucer and glue to bottom of cup. Set aside to dry. Trim saucer with rickrack to resemble collar. Glue yarn in short loops to top of cup for hair. Glue rickrack over edges of yarn loops. Glue small square of felt for eyes on each side of cup handle (nose). Add buttons or sequins in center of eyes. Add "mouth" of red felt. Use as small planter.

Shoe Buffer

If, at this point, your husband is starting to notice that you have taken over his dresser drawers for storage of your treasures and/or he discovers that his galoshes are filled with empty match boxes (after all, he only wears them in heavy snow), it is time to finish your current project and create something just for him. Not only will he be amazed at your ingenuity, he will probably offer you his old Army duffel bag for future storage of your finds. This is a supreme sacrifice on his part as this memento is the only item he has ever saved.

POTPOURRI 91

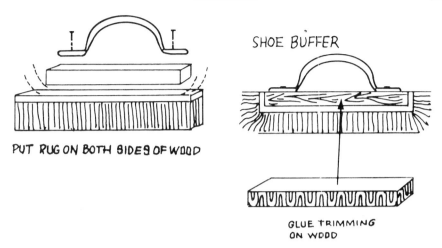

Materials needed:
  discarded child's wooden block
  grip-type cabinet pull
  white glue
  carpet remnant
  scrap of braid-type trim

Attach cabinet pull (handle) to center of block, using screwdriver. Cut piece of rug remnant to size of block, including side measurements. Glue into place on bottom and ends of block. Glue strip of braid trim around edge of block where wood meets rug.

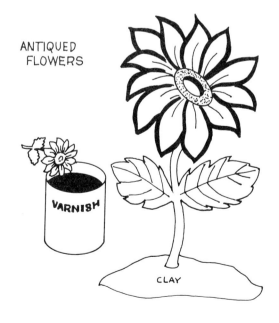

92   MAKING TREASURES FROM TRASH

Antiqued Flowers

The results of this "antiquing" process are unbelievable. This is a simple way of recycling all those old plastic flowers you have been saving.

Materials needed:
  plastic flowers
  ½ pint turpentine
  ½ pint walnut or oak varnish

Mix turpentine and varnish into any old container deep enough for dipping plastic flower (an empty coffee can is perfect). Dip each flower into mixture, holding upside down by stem end, making sure all petals are submerged. Shake excess mixture from flower as you withdraw from container. Dry upright, sticking stems into clay, or hang by stems on clothes line. Experiment with different colors of varnish. Each produces a unique type of flower. All are foolproof, and the results are spectacular.

Blooming Button Pot

A clever gift, and one which is always welcome in a home, is this button box.

Materials needed:
  plastic flower pot
  fabric scrap, approximately ⅛ yard
  white glue
  small yarn pom-pom
  pincushion stuffing (shredded foam, old hose, etc.)

Cut 1" strip of fabric to go around widest part of flower pot (near top). Add ½" to length of strip for overlap. Glue into place. Cut wider strip of fabric for lower portion of pot, adding ½" for overlap. Glue into place, smoothing out any wrinkles as you work. Cut 2 circles of fabric the diameter of flower pot top, adding ½" for seam allowance. With right sides together, sew circles together, leaving small opening to insert stuffing. Turn right side out and fill with stuffing. Hand stitch opening. Attach yarn pom-pom to center of pincushion. A button may be used in place of pom-pom. Place pincushion in pot.

Spray-Can Covers

Is there anyone out there who has not, at least once, mistaken hair spray for furniture polish spray? Just recently, the top of my dresser was sprayed with deodorant in error. The dresser didn't shine, but it didn't perspire either! So despite this aerosol age we live in, I have devised a way to "dress up" some of the essential spray products used regularly. In addition to being pretty to look at, these covers help identify their contents.

Materials needed:
   lightweight cardboard (like those that come with laundered shirts)
   wallcovering samples
   assorted trimmings (artificial flowers, beads, jewels, etc.)
   stapler
   white glue

Measure spray-can from top to bottom (not including cap), also diameter of can. Cut cardboard to these measurements, adding 1" to diameter measurement for overlap. Cut piece of wallcovering sample same size as cardboard and glue to cardboard. Allow to dry. Staple covered cardboard together at top, center and bottom of seam line. Trim as desired. Slip cover over spray can.

Bamboo Wall Hangings

Every family member can "do his own thing" in creating these colorful wall hangings. Imagination provides an endless variety of designs, or use conventional objects such as flowers, animals, etc. A simple pattern can be followed by using a child's coloring book.

Materials needed:
   burlap material—amount needed will depend on size of wall hanging desired
   scraps of fabric or felt, yarn, ribbon, buttons, etc.
   white glue
   2 bamboo rods, each 4" longer than size of wall hanging desired
   Bamboo poles are used to roll carpeting and similar floor coverings. Any carpet dealer will gladly give you these bamboo poles just for the asking.

Cut burlap to desired size, allowing an extra inch for hem on top and bottom. Turn back hems and glue or sew into place. Fringe outer edges of burlap. Use hairpin to fringe, pulling threads out one at a time until 1" fringe is pulled on both sides. Save threads to make

POTPOURRI 95

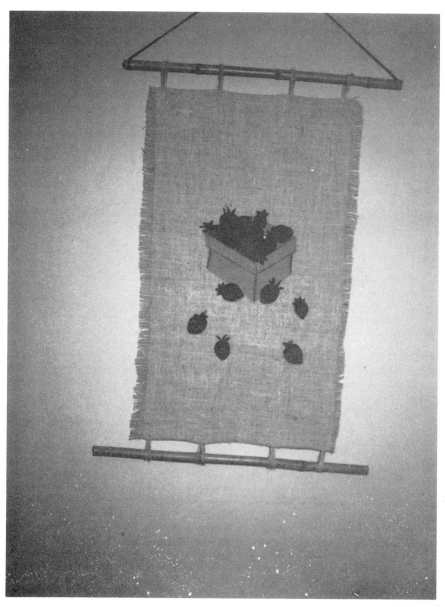

loops for hanging. Draw desired pattern on burlap or trace a picture from a coloring book. Cut pattern from fabric and glue each pattern piece into place on burlap. Use yarn or other trimming for details. Make 8 loops from burlap threads, each 6" long, and sew to top and bottom hems at back, spacing each loop at even intervals. Slip bamboo rod through loops. Cut slits in top bamboo rod for inserting cord used to hang on wall. Hang in prominent spot and wait for the compliments.

STAMP BOX

Stamp Boxes

After several sticky mishaps with my postage stamps, formerly stored in my purse (always near tissues), I decided to make a personal stamp storage box small enough to keep in my purse, yet handy to reach for, when needed. Make several and give to children going away to camp, a distant school, etc.

Materials needed:
  small match boxes
  covering of your choice: felt, wallcovering, canceled stamps
  white glue

Remove box top from sliding bottom. Cut covering to fit over box top. Glue into place. If using felt, use magic marker to draw "canceled stamp." If using old stamps, overlap each one in random pattern.

Drapery Ring Wall Accent

These mini-frames make ideal boutique items for bazaars and housewarming gifts.

Materials needed:
  wooden drapery rings
  lightweight cardboard

DRAPERY RING WALL ACCENT

fabric scrap
small dried flowers
narrow ribbon
white glue

Place drapery ring on cardboard and trace inner circle dimension. Cut cardboard circle out, place on fabric scrap and cut circle out of fabric. Glue fabric to cardboard. Choose 3 or 4 small dried flowers and arrange in center of circle to resemble mini-bouquet. Glue into place. Make small bow out of ribbon and glue over flower stems as shown. Apply thin strip of glue to outer edges of circle and press into place at back of drapery ring, having hanging hook on top. Make small bow out of ribbon and glue onto hanging hook. Several of these can be grouped into interesting wall patterns.

Clothespin Trivet

Since the clothes dryer has replaced the clothes line, what do you do with all those wooden clothespins you have saved all these years? I tried giving them to my children for playing "pick-up sticks," but they claimed they needed Jolly Green Giant hands to help pick them up. So we made trivets.

Materials needed:
  wooden clothespins
  white glue
  optional—paint or clear varnish

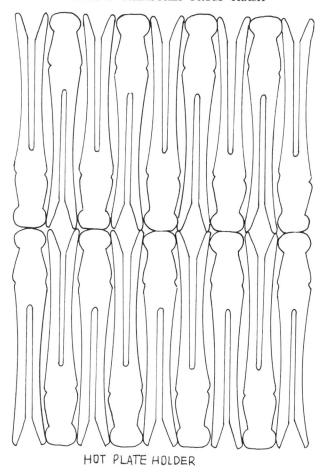
HOT PLATE HOLDER

Lay out clothespins into pattern. There are countless combinations to use, circular, square, etc. After you have played around a bit and decided on a pattern, glue clothespins to one another at sides. Set aside to dry. If desired, paint trivet color of your choice (good for really old, weather-beaten pins) or cover with clear varnish. If newer pins are used, nothing else need be done after gluing together.

Family Tree

A family tree can be made using wooden drapery rings, substituting family member pictures for fabric-flower centers. Simply cut picture into center circle size and glue into place at back of drapery ring. Join "tree" with narrow ribbon through hanging hooks. Make bow at top ring hook.

FAMILY TREE

Spice Sachet Ball

Among the many so-called useless items discovered in my "junk" drawer was a metal tea strainer. I do not remember times prior to the invention of the tea bag, so I guess this item was part of a legacy left by a non-collector of valuables. After making the first sachet ball, everyone who saw it wanted one, so I ended up buying a dozen

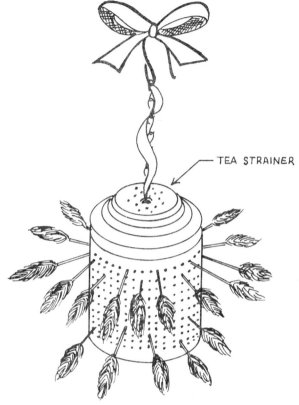

SPICE SACHET BALL

strainers. Don't even ask about the clerk's reaction to that purchase!

Materials needed:
    metal tea strainer (bought at variety store)
    whole cloves, bit of cinnamon stick
    small plastic fruits, or flowers
    narrow ribbon

Remove top from strainer and fill with cloves and small pieces of cinnamon stick. Replace top. Carefully insert stems of fruits or flowers into strainer holes, until strainer is completely covered. Insert ribbon through chain at top of strainer and make bow. Hang in closets or anywhere you would like a refreshing fragrance.

Mini-Vases

Tops of lipstick tubes make mini-vases for an endless variety of artificial flowers. My personal favorite is the tiny dried straw flower, but try your hand at different varieties. Lipstick tubes come in such a wide range of pretty designs and colors, no other decoration or paint is necessary.

Materials needed:
  lipstick tube (top)
  small amount of modeling clay
  flowers of your choice

Measure length of lipstick top so that completed arrangement will allow flower and small amount of stem to show above top edge of tube. Insert flowers into small lump of clay. Drop into tube and press clay firmly into place, using toothpick or similar item. Take a mini-vase along on a hospital visit to a patient. Sure to cheer.

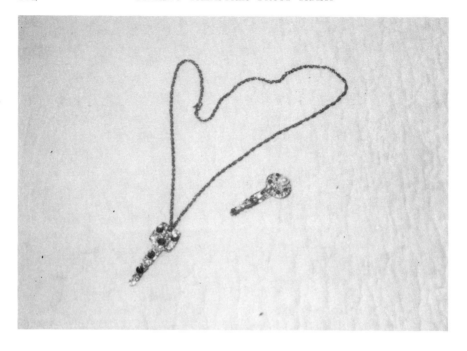

Key Jewelry

The front door lock has been replaced 4 times since we have lived here. The first time we needed a new lock was due to a broken-off key left in the keyhole (my son, Samson). The next time we had to replace the lock was due to some bubble gum left to harden in the keyhole (my son, the mischief maker). The third occasion for a new lock was due to me; I chain-locked the door, and we had to crow-bar the door to get in (me, the dummy). We now have our 4th lock in place, and own 14 useless keys.

Pin

Materials needed:
  old key
  pin back (found in variety store)
  white glue
  model paint; gold or silver or color of your choice
  jewels (beads, gems, etc.)

Paint both sides of key. When paint has dried, glue pin back to underside of key. Position "jewel" over hole in key and glue in place. If desired, smaller jewels can be glued on key.

Necklace

Paint key desired color. Glue small "jewels" or pearls on key. Use discarded chain to draw through hole in key, or for a novel effect, use leather shoe lace (type used for work boots).

These key jewelry items make clever "Door Prizes" when placed into small match boxes which have been painted and decorated to resemble doors.

Gingham Garden Flowers

Fabric scraps from your sewing basket are used to make these delightful posies. Use single large flowers or an array of smaller ones to add a country-fresh touch to any room.

Materials needed:
  fabric scraps (gingham, calico)
  pipe cleaners
  white glue

Use one pipe cleaner per petal. Wrap each pipe cleaner around a juice glass or small bottle and twist ends (this forms all petals same

POTPOURRI 105

size). Use 4 or 5 petals for each flower and make leaves the same way, using 1 or 2 per flower. Apply glue to one side of pipe cleaner circle and place glue side down on fabric. Allow to dry, then cut fabric around outer edge of pipe cleaner. Bend petals and leaves into desired shapes or leave round, bending stem portion down. Form

MAKING TREASURES FROM TRASH

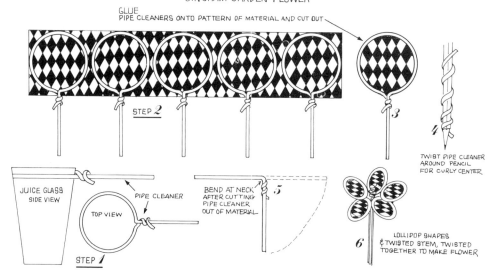

flower center by wrapping pipe cleaner around a pencil. Slip the spiral from pencil and position petals around it. Wrap one pipe cleaner end around entire base of flower. Attach leaves to flower stem and twist pipe cleaner around to secure.

Decorated Soap

Jiffy and inexpensive gift items can be made from ordinary bars of soap and bits of odds and ends.

POTPOURRI 107

Floral Soap

Materials needed:
  bar of soap, oval shape
  small artificial flowers
  scraps of trimming or narrow ribbon
  small beads, pearls, sequins
  straight pins

Remove any stems from flowers. Insert straight pin into hole of bead, then through center of flower and push into top surface of soap. Attach leaves the same way. Cover the entire top surface of soap. Position ribbon around edge of soap and pin into place with beads at both ends. Make stand for floral soap by inserting pin through medium-sized bead, then pushing into bottom of soap at 4 corners. For gift-giving, place into florist's corsage box with lots of tissue paper.

Soap Candle

Materials needed:
  bar of soap

wash cloth
two artificial flowers
scrap of yellow or orange felt
piece of netting
straight pins
narrow ribbon

Fold washcloth in half and roll tightly. Secure ends with pins pushed into roll. Cut felt into shape of "flame," as illustrated. Place flame into center of rolled washcloth. Secure roll to center of soap with straight pins. Cut 12" square of netting, and place soap in center. With narrow ribbon, tie netting around soap at base of "candle." Insert pins into centers of flowers and attach to soap as shown.

Soap Fish

Materials needed:
  oval-shaped soap

POTPOURRI 109

netting
small pearls and sequins
straight pins
narrow ribbon

Cut netting into a 10" square. Place soap into center and tie netting at one end of soap with narrow ribbon. Insert straight pin through 2 small pearls and 1 small sequin, then push into top edge of soap. Place 4 of these beaded pins in a row. Insert pins through pearl and sequin and push into sides of soap to form triangle. For "eyes," insert pin through pearl, small sequin, then large sequin and push into sides of soap.

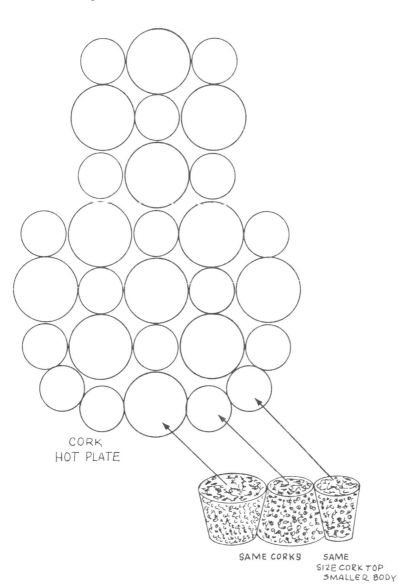

CORK HOT PLATE

SAME CORKS   SAME SIZE CORK TOP SMALLER BODY

Cork Trivet

Materials needed:
  corks, all one size
  white glue

Glue corks one to another, placing one cork up and one cork down. Continue gluing until desired size trivet is made. Trivet may be square, round, or any shape you desire. To make an unconventional shape, it is easy to follow shape if first drawn on piece of paper. Place corks within drawn pattern and glue together until form is completed.

HAT-FELT, COLLAR-BUTTON   HAT-FELT, ARMS, TIE

Family Name Plates

I don't think there are any of us who do not have several odd dinnerware pieces tucked away in our cabinets. Not enough for even one place setting, yet enough to hold onto for some unforeseen use. Now is the time to put them on display and decorate a wall at the same time.

Materials needed:
  odd dinnerware pieces
  felt scraps

fabric scraps and rickrack
white glue
model paint

Following illustration, cut pattern pieces out of fabric and felt. Use pink felt for face and cut features (eyes, mouth, hair) out of appropriately colored felt scraps. Glue face onto plate, then glue features in place. Position fabric pieces and glue onto plate as shown. Edge "clothing" with rickrack. Paint (use black or match color scheme of room in which plates are to be displayed) border around plate. If dinnerware has an embossed border, just follow design with paint. Paint family member name on plate as shown.

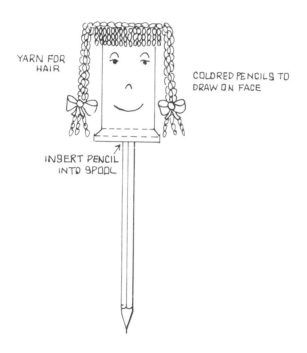

Spool Heads

Use these adorable "spool heads" for party favors and as gifts for small friends.

Materials needed:
  wooden spool
  colored pencils or magic markers

yarn scraps
white glue

Draw eyes, nose and mouth with colored pencils on spool. Use yarn for hair. For the girls, make braids using 18 strands of yarn approximately 8" long. Tie ends with ribbon and glue across top of spool. Glue short yarn loops over top front of spool for bangs. For the boys, glue short yarn loops on entire top of spool. Insert pencil through bottom hole.

77-11959   745.5
            DAR

77-11959   745.5
         53398   DAR

AUTHOR
Darman, Marion

TITLE
Making treasure from trash

| DATE DUE | BORROWER'S NAME | ROOM |

745.5
DAR   Darman, Marion

        Making treasures from
        trash

        53398

LOURDES HIGH SCHOOL LIBRARY
4034 WEST 56th STREET
CHICAGO 29 ILLINOIS